Unit 2
From Matter to Organisms

Copyright © 2020 by Discovery Education, Inc. All rights reserved. No part of this work may be reproduced, distributed, or transmitted in any form or by any means, or stored in a retrieval or database system, without the prior written permission of Discovery Education, Inc.

NGSS is a registered trademark of Achieve. Neither Achieve nor the lead states and partners that developed the Next Generation Science Standards were involved in the production of this product, and do not endorse it.

To obtain permission(s) or for inquiries, submit a request to:

Discovery Education, Inc.
4350 Congress Street, Suite 700
Charlotte, NC 28209
800-323-9084
Education_Info@DiscoveryEd.com

ISBN 13: 978-1-68220-804-5

Printed in the United States of America.

1 2 3 4 5 LBC 28 27 26 25 24

Acknowledgments

Acknowledgment is given to photographers, artists, and agents for permission to feature their copyrighted material.

Cover and inside cover art: Kevin Wells Photography / Shutterstock.com

Table of Contents

Unit 2: From Matter to Organisms

Letter to the Parent/Guardian vi

Unit Overview 1

   Anchor Phenomenon: Food Chains and Food Webs 2

Unit Project Preview: Build a Mini-Ecosystem 4

Concept 2.1 Plant Needs

Concept Overview 6

   Wonder 8

      Investigative Phenomenon: Tree Needs 10

   Learn 18

   Share 40

Concept 2.2 Matter Flow in Ecosystems

Concept Overview 48

   Wonder 50

      Investigative Phenomenon: How Hawks Find Food 52

   Learn 60

   Share 94

Concept 2.3 Energy Flow in Ecosystems

Concept Overview 104

    Wonder 106

        Investigative Phenomenon: The Sun and Plants 108

    Learn 114

    Share 130

Unit Wrap-Up

Unit Project: Build a Mini-Ecosystem 138

Grade 5 Resources

Bubble Map R3

Safety in the Science Classroom R4

Vocabulary Flash Cards R7

Glossary R27

Index R58

Unit 2: From Matter to Organisms

Dear Parent/Guardian,

This year, your student will be using Science Techbook™, a comprehensive science program developed by the educators and designers at Discovery Education and written to the Next Generation Science Standards (NGSS). The NGSS expect students to act and think like scientists and engineers, to ask questions about the world around them, and to solve real-world problems through the application of critical thinking across the domains of science (Life Science, Earth and Space Science, Physical Science).

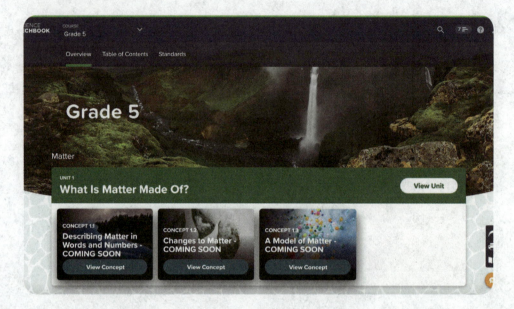

Science Techbook is an innovative program that helps your student master key scientific concepts. Students engage with interactive science materials to analyze and interpret data, think critically, solve problems, and make connections across science disciplines. Science Techbook includes dynamic content, videos, digital tools, Hands-On Activities and labs, and game-like activities that inspire and motivate scientific learning and curiosity.

You and your child can access the resource by signing in to www.discoveryeducation.com. You can view your child's progress in the course by selecting the Assignment button.

Science Techbook is divided into units, and each unit is divided into concepts. Each concept has three sections: Wonder, Learn, and Share.

Units and Concepts Students begin to consider the connections across fields of science to understand, analyze, and describe real-world phenomena.

Wonder Students activate their prior knowledge of a concept's essential ideas and begin making connections to a real-world phenomenon and the **Can You Explain?** question.

Learn Students dive deeper into how real-world science phenomenon works through critical reading of the Core Interactive Text. Students also build their learning through Hands-On Activities and interactives focused on the learning goals.

Share Students share their learning with their teacher and classmates using evidence they have gathered and analyzed during Learn. Students connect their learning with STEM careers and problem-solving skills.

Within this Student Edition, you'll find QR codes and quick codes that take you and your student to a corresponding section of Science Techbook online. To use the QR codes, you'll need to download a free QR reader. Readers are available for phones, tablets, laptops, desktops, and other devices. Most use the device's camera, but there are some that scan documents that are on your screen.

For resources in Science Techbook, you'll need to sign in with your student's username and password the first time you access a QR code. After that, you won't need to sign in again, unless you log out or remain inactive for too long.

We encourage you to support your student in using the print and online interactive materials in Science Techbook, on any device. Together, may you and your student enjoy a fantastic year of science!

Sincerely,

The Discovery Education Science Team

Unit 2
From Matter to Organisms

Get Started

Food Chains and Food Webs

The flow of materials and energy through an ecosystem is called a food chain. A network of food chains is called a food web. An ecosystem consists of living and nonliving things. At the end of this unit, you will be able to develop a model of an ecosystem that can be used to explain how different organisms are connected in an ecosystem.

Quick Code: us5255s

Food Chains

Think About It

Look at the photograph. **Think** about the following questions:

- What matter do plants need to grow?
- How does matter move within an ecosystem?
- How does energy move within an ecosystem?

Ducks in Their Ecosystem

Unit 2: From Matter to Organisms

Unit Project Preview

 Solve Problems Like a Scientist

Unit Project: Build a Mini-Ecosystem

In this project, you will use what you know about ecosystems and energy flow to build your own mini-ecosystem.

Quick Code: us5256s

Mini-Ecosystems

SEP Developing and Using Models
CCC Energy and Matter

Ask Questions About the Problem

You are going to design and create a mini-ecosystem using what you know about relationships and energy flow in ecosystems. **Write** some questions you can ask to learn more about the problem. As you learn about how matter and energy flow through ecosystems in this unit, **write** down the answers to your questions.

Plant Needs

CONCEPT 2.1

Student Objectives

By the end of this lesson:

- [ ] I can use evidence to argue that plants get most of the materials they need to grow from air and water.

- [ ] I can develop a model of how matter moves through plants and of plant processes that interact with each other and depend on each other.

Key Vocabulary

- [ ] air
- [ ] light energy
- [ ] nutrient
- [ ] oxygen
- [ ] plant
- [ ] stem
- [ ] stomata
- [ ] survive
- [ ] system
- [ ] water

Quick Code: us5258s

Activity 1
Can You Explain?

How do plants use water, air, and light to fulfill their basic needs?

_____
_____
_____
_____
_____
_____
_____
_____
_____
_____
_____
_____

Quick Code:
us5259s

2.1 | Wonder How do plants use water, air, and light to fulfill their basic needs?

Activity 2
Ask Questions Like a Scientist

Quick Code: us5260s

Tree Needs

Look at the photograph. Then, **answer** the questions that follow.

Let's Investigate Tree Needs

SEP Asking Questions and Defining Problems
SEP Developing and Using Models

My Questions

Write down questions you have about the student planting a tree.

_____

_____

_____

Draw a model of how a plant meets its needs. Your model can be words, pictures, symbols, or any combination of these choices.

My Model

Concept 2.1: Plant Needs | 11

2.1 | Wonder
How do plants use water, air, and light to fulfill their basic needs?

Activity 3
Observe Like a Scientist

Growing

Watch the video. **Complete** the Change over Time graphic organizer to record how a seedling changes as it grows to a mature plant.

Quick Code: us5261s

Growing

Before	After

Changes

Activity 4

Observe Like a Scientist

Quick Code: us5262s

Water in the Desert

Look at the images of the Sonoran Desert before and after rainfall. Pay close attention to the plants and ground.

Desert before Rain

Desert after Rain

Concept 2.1: Plant Needs | 13

2.1 | Wonder
How do plants use water, air, and light to fulfill their basic needs?

Describe the similarities and differences between the desert before and after the rain.

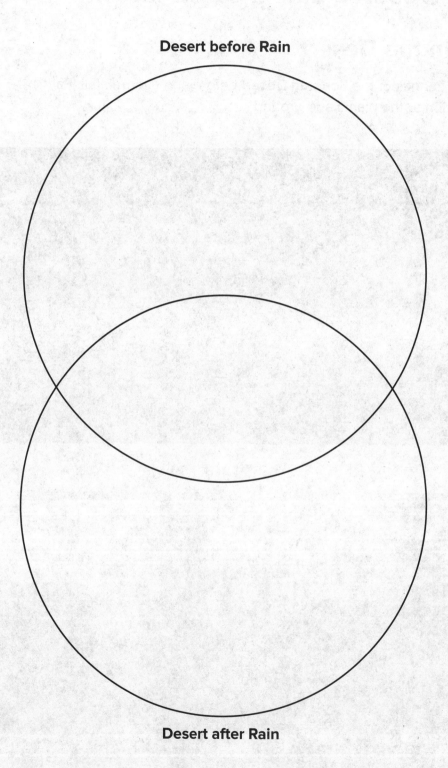

Desert before Rain

Desert after Rain

14

Activity 5
Evaluate Like a Scientist

What Do You Already Know About Plant Needs?

Quick Code: us5263s

Plants and Animals

Think about what animals and plants need to live and grow. Then, **answer** the questions.

What do plants need to live and grow?

_____
_____
_____
_____

How are the needs of plants similar to those of humans?

_____
_____
_____
_____

How are they different?

_____
_____
_____
_____

2.1 | Wonder
How do plants use water, air, and light to fulfill their basic needs?

Plant Needs

Think about what plants need to live and grow. **Label** each item listed as "Basic Plant Need" or "Not Basic Plant Need."

Soil _____

Water _____

Sugar _____

Oxygen _____

A forest _____

Carbon dioxide _____

Plants and Food

Read the questions. Then, **write** your answers in the space provided.

How do plants get their food?

_____
_____
_____

How do the roots, stems, and leaves of the plant each help it get food?

_____
_____
_____

Concept 2.1: Plant Needs

2.1 | Learn How do plants use water, air, and light to fulfill their basic needs?

How Does a Plant Get the Materials It Needs to Survive and Grow?

Activity 6

Investigate Like a Scientist

Quick Code: us5264s

Hands-On Investigation: Do Plants Need Soil?

In this activity, you will **investigate** whether plants need soil to grow. You will **germinate** seeds in wet paper towels, **measure** their growth, and then **compare** their growth to the growth of the seed potted in soil.

Make a Prediction

Do you agree with the claim: *Plants can grow without soil*? **Circle** your choice in the Claim-Evidence-Reasoning table, and then give evidence and reasoning to support your choice.

Claim

Plants can grow without soil.
Agree or Disagree?

 SEP Planning and Carrying Out Investigations
 SEP Engaging in Argument from Evidence

What materials do you need? (per group)

- Plastic cup, 9 oz
- Soil, potting
- Paper towels
- Seeds, lima beans
- Plastic zipper bags
- Water
- Pen or marker
- Metric ruler
- Balance, triple beam
- Lettuce or similar small plants (optional)

Evidence

Reasoning

Concept 2.1: Plant Needs

2.1 | Learn — How do plants use water, air, and light to fulfill their basic needs?

What Will You Do?

1. Use the water in the cup to wet the paper towel.
2. Place three of your bean seeds in the top half of the paper towel. Fold the bottom half of the towel up so that it covers the beans. Place the paper towels inside the plastic zip bag and seal it.
3. Plant the fourth bean seed in the cup of soil. Water the seed.
4. Label the beans with your name, then place the plastic zip bag and the soil cup in a place where they can get sunlight.
5. Check the growth of your seeds over the next several days. Dampen the paper towel and water the soil.
6. Create a table to record your data. Measure the growth of each seed and record the measurements labeled by the date and whether the seed being measured is germinating in the towel or in the soil cup.

Data Table

Think About the Activity

How much did the beans that were placed in the paper towels grow? How did they compare with the control bean planted in soil?

_____
_____
_____
_____

Did the growth of the beans, both in soil and in paper towels, match your hypothesis? If not, how was it different?

_____
_____
_____

Based on your observations, do seeds need soil to grow? Can plants grow entirely without soil? If so, will they grow better in soil?

_____
_____
_____
_____
_____
_____
_____

Concept 2.1: Plant Needs | 21

2.1 | Learn How do plants use water, air, and light to fulfill their basic needs?

Activity 7
Investigate Like a Scientist

Quick Code: us5265s

Hands-On Investigation: Sunlight: A Basic Need

In this investigation, you will **perform** an experiment to **look** for any difference in how plants grow in the light and in the dark. You will then **compare** and **contrast** your observations with your classmates. You will **set up** the activity today and complete the activity later in this concept.

Make a Prediction

Develop a claim about what you think will happen to the plants.

What do you predict will happen to the plant in the light?

_____
_____
_____
_____

What do you predict will happen to the plant in the dark?

_____
_____
_____
_____

| SEP | Planning and Carrying Out Investigations |
| SEP | Engaging in Argument from Evidence |

What materials do you need? (per group)

- Plastic cups, 9 oz
- Soil, potting
- Water
- Black permanent marker

What Will You Do?

1. Use the permanent marker to write your names on the cups and label the cups A and B.

2. Add soil to your cups. Place the bean seeds on the soil, one per cup, and cover the seeds with about 2 centimeters of soil. Add the same amount of water to each cup to moisten the soil.

3. Place cup A where it will receive light and place cup B in the dark.

4. Use the checklist to care for your seeds/plants. Record height and color of plant for five days once the plant can be seen growing above the surface of the soil.

2.1 | Learn
How do plants use water, air, and light to fulfill their basic needs?

Data Table for Plant Growth

Checking the Basic Needs of Plants	Cup A (light)	Cup B (darkness)
Check the plant every day.		
Make sure the plant has air.		
Water the plant when the soil feels dry.		
Date new plant was seen growing above the surface of the oil. This will became Day 1.		
Color of new plant		
Height and color of new plant on Day 2		
Height and color of new plant on Day 3		
Height and color of new plant on Day 4		
Height and color of new plant on Day 5		
Results:		
Plant that is the tallest after Day 5		
Plant that seems more healthy after Day 5		
Color of plants after Day 5		

After collecting your data over a period of several class periods, you will analyze your data. You should compare and contrast your observations with your classmates.

Think About the Activity

What are the basic needs of plants?

_____

What happened to the plant in the light?

_____

_____

What happened to the plant in the dark?

_____

_____

Explain why light is important to plant growth. Include sketches to support your conclusions.

Concept 2.1: Plant Needs | 25

Activity 8
Analyze Like a Scientist

Plant Structure

Quick Code: us5266s

Read the following text. As you read, **draw** the different plant parts in the boxes on the Summary Frames on the next pages. **Write** about the parts on the lines under the boxes.

Plant Structure

All living things have basic needs that they must meet to **survive**. For example, you need **water**, **air**, and food to live. Plants are living things, too. Like people, all plants need water and air to survive. Of course, plants and humans are very different. You get your food from plants and animals, but plants use sunlight to make their own food from air and water.

Some Native Flowers and Plants of Granada

Plants have ways to capture and move the water they need to survive. A **plant**'s roots absorb water from the soil and carry the water to the rest of the plant. Roots also carry nutrients from the soil to the plant. Water and **nutrients** move up and down a plant's **stem** through tubes called vessels. Smaller vessels connect the stem to leaves. This **system** helps feed and water all the parts of the plant. The air that plants need moves directly into leaves through tiny openings called **stomata**. Leaves also collect sunlight.

CCC Structure and Function

Concept 2.1: Plant Needs

Activity 9
Observe Like a Scientist

Parts of a Plant

Watch the video. As you watch, **record** any new information about different plant parts on the Summary Frames from the previous activity.

Quick Code: us5267s

Parts of a Plant

CCC Structure and Function

Concept 2.1: Plant Needs | 29

2.1 | Learn How do plants use water, air, and light to fulfill their basic needs?

Activity 10
Investigate Like a Scientist

Hands-On Investigation: Up the Stem

Quick Code: us5268s

In this investigation, you will **observe** how plants move water. You will use your observations to **compare** how different types of plant stems move water.

Make a Prediction

Develop a claim about what you think will happen to the celery stalks when you place them in the cup of colored water.

_____

_____

_____

_____

What materials do you need? (per group)

- Celery stalk or flower stem
- Plastic cup, 9 oz
- Food coloring
- Water
- Scissors
- Hand lens

CCC Structure and Function

What Will You Do?

1. Select a stem. Examine your stem closely. Record observations about how the stem looks and how it feels.

2. Put food coloring in the cup of water, snip about an inch off the bottom of the stalk or stem, and place it in the water.

3. Leave the stalks and stems in the water cups and set aside where they will not be disturbed until the next day.

4. Observe the stalks. Record your observations and compare the actual outcome with your prediction.

Before	After

Changes

Concept 2.1: Plant Needs | 31

2.1 | Learn
How do plants use water, air, and light to fulfill their basic needs?

Think About the Activity

How did your predictions about the outcome of the investigation differ from your observations?

_____

What can you conclude about the different stems' ability to transport water?

_____

Activity 11
Observe Like a Scientist

Basic Needs

Complete the interactive to learn about the basic needs of a plant. **Observe** how the different parts of a plant work to fulfill these basic needs. As you work, **record** the functions of each plant part in the table.

Quick Code: us5269s

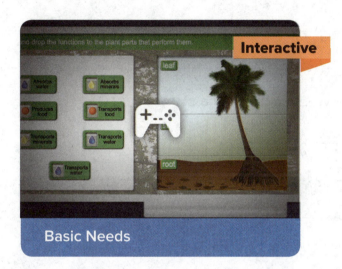

Basic Needs

Plant Part	Functions
Leaves	
Stems	
Roots	

CCC Structure and Function

Concept 2.1: Plant Needs | 33

2.1 | Learn
How do plants use water, air, and light to fulfill their basic needs?

Now, **use** your observations to **answer** the questions.

How does the stem help a plant meet its basic needs?

_____
_____
_____
_____

How does water move into and out of a plant?

_____
_____
_____
_____

How will a plant be affected if animals eat all its leaves?

_____
_____
_____
_____
_____

Activity 12
Evaluate Like a Scientist

Obtaining Materials

Quick Code: us5270s

Write the words in the boxes to show how a leaf gets the materials it needs to grow. You may not need all of the words.

| air | soil | sunlight | water |

Obtaining Materials

leaf

CCC Structure and Function

Concept 2.1: Plant Needs | 35

How Does a Plant Use Materials from Air and Water to Grow?

Activity 13
Analyze Like a Scientist

Plant Food

Read the text describing the steps a plant takes to convert energy into food. **Number** each step in the paragraphs that follow. Then, **compare** and **discuss** your numbering with a partner. Once you and your partner agree, **write** the steps in the table.

Quick Code: us5271s

Plant Food

Plants combine water with a gas in the air to make a sugar called glucose. Plant cells use this glucose for food. This process happens in a plant's leaves. Light from the sun provides the energy needed for this food-making process. Remember that energy can be transformed from one form to another. During this process, **light energy** from sunlight is transformed into chemical energy that is found in glucose.

Special structures called vessels help move glucose from the leaves to the other parts of the plants. Plant cells use glucose as a source of energy to live and grow. As they use glucose, they release **oxygen** and water into the air. These materials are considered waste products of the plants. Other living things, such as animals, depend on the oxygen that plants release during this process of food production. Plants also have structures that take in water and nutrients from the soil, and move them to other parts of the plant.

Step Number	Step Description

CCC Structure and Function

2.1 | Learn — How do plants use water, air, and light to fulfill their basic needs?

Activity 14
Observe Like a Scientist

Quick Code: us5272s

Leaves and Food Production

Watch the video. **Look** for information about how leaves make food.

Leaves and Food Production

Talk Together

Now, listen to the statements your teacher reads and talk together about which statement is a lie. How do you know it is not true?

CCC Structure and Function

Activity 15

Evaluate Like a Scientist

Quick Code: us5273s

Leaf or Factory?

A leaf can be considered a system. Systems have inputs and outputs. A factory can also be considered a system. **Write** Leaf or Factory to match each system to its output or input.

This system makes food.

This system needs metal and rubber.

This system gets energy from sunlight.

This system needs air and water.

This system makes cars.

This system gets energy from electricity.

Draw pictures of a leaf and a factory. Include the inputs and outputs of each system.

Leaf	Factory

SEP Developing and Using Models

Concept 2.1: Plant Needs | 39

2.1 | Share How do plants use water, air, and light to fulfill their basic needs?

Activity 16

Record Evidence Like a Scientist

Quick Code: us5274s

Tree Needs

Now that you have learned about plant needs, look again at the image Tree Needs. You first saw this in Wonder.

Let's Investigate Tree Needs

Talk Together

How can you describe Tree Needs now?
How is your explanation different from before?

SEP Constructing Explanations and Designing Solutions

Look at the Can You Explain? question. You first read this question at the beginning of the lesson.

> **Can You Explain?**
>
> How do plants use water, air, and light to fulfill their basic needs?

Now, you will use your new ideas about Tree Needs to answer a question.

1. **Choose** a question. You can use the Can You Explain? question or one of your own. You can also use one of the questions that you wrote at the beginning of the lesson.

   My Question

2. Then, use the Claim-Evidence-Reasoning graphic organizer on the next page to answer the question.

Concept 2.1: Plant Needs | 41

2.1 | Share
How do plants use water, air, and light to fulfill their basic needs?

Work with a partner to **write** a claim and **gather** evidence to answer the Can You Explain? question. Brainstorm what science words you want to use in your reasoning. Use the table below to **record** your responses.

My claim (the answer to the question)

Evidence I have found (record all evidence you gathered from video, reading, interactives, and Hands-On Investigations)

Reasoning: My claim is true because:

STEM in Action

Activity 17
Analyze Like a Scientist

Farmers Growing Plants: Irrigation and Sunlight

Quick Code: us5275s

Read the text. **Look** at the images. **Discuss** one of the irrigation methods with your group. **Write** the pros and cons of your irrigation method in the table that follows.

Farmers Growing Plants: Irrigation and Sunlight

One of the basic needs of plants is water. Farmers grow many plants across large areas. They cannot always rely on rainfall to water their plants. Because of this, they have developed many different ways to get water to their plants. These photos show a number of irrigation systems. You can see irrigation sprinklers, an irrigation canal, and an irrigation pump truck.

Irrigation Sprinklers

SEP Engaging in Argument from Evidence

Concept 2.1: Plant Needs | 43

Farmers Growing Plants: Irrigation and Sunlight *cont'd*

Irrigation Canal

Irrigation Pump Truck

We know that water is a basic need of plants. The images show many ways that farmers work to get water to their crops. Another basic need of plants is sunlight. Look at the photo of the plant's leaves in the sunlight. Leaves use sunlight for the process of photosynthesis. Sunlight is the energy source. The amount of sunlight a plant receives affects how much the plant grows. You can see why this is very important to farmers. Sunlight directly affects the size of their crop.

Leaf in Sun

Irrigation Method:

Pros:	Cons:

Concept 2.1: Plant Needs

2.1 | Share
How do plants use water, air, and light to fulfill their basic needs?

Activity 18
Evaluate Like a Scientist

Quick Code: us5276s

Review: Plant Needs

Think about what you have read and seen. What did you learn?

Write down some key ideas you have learned. **Review** your notes with a partner. Your teacher may also have you take a practice test.

_____
_____
_____
_____
_____
_____
_____
_____
_____
_____
_____
_____
_____
_____

SEP Obtaining, Evaluating, and Communicating Information

Talk Together

Think about what you saw in Get Started. Use your new ideas about plant needs to discuss how organisms in an ecosystem are connected.

CONCEPT 2.2

Matter Flow in Ecosystems

Student Objectives

By the end of this lesson:

- [ ] I can develop a model of how matter is conserved as it moves through an ecosystem.
- [ ] I can develop a model based on evidence of how animals feed on each other.
- [ ] I can explain how decomposers help move matter in the environment.

Key Vocabulary

- [ ] absorb
- [ ] consumer
- [ ] cycle
- [ ] decomposer
- [ ] ecosystems
- [ ] environment
- [ ] food chain
- [ ] food web
- [ ] interact
- [ ] prey
- [ ] producer
- [ ] recycle
- [ ] shelter
- [ ] sun

Quick Code: us5278s

Concept 2.2: Matter Flow in Ecosystems

Activity 1
Can You Explain?

How does matter flow through an ecosystem?

_____
_____
_____
_____
_____
_____
_____
_____
_____
_____
_____

Quick Code:
us5279s

2.2 | Wonder How does matter flow through an ecosystem?

Activity 2
Ask Questions Like a Scientist

How Hawks Find Food

Quick Code: us5280s

Look at the photograph. Then, **answer** the questions that follow. **Record** your answers in the space provided.

Let's Investigate How Hawks Find Food

CCC Patterns

Your Questions

What questions would you ask to find out about how a hawk survives in its environment?

_____

_____

_____

Draw a model of how the hawk interacts with the environment. You can use words, images, and/or symbols.

My Model

Concept 2.2: Matter Flow in Ecosystems | 53

Activity 3
Analyze Like a Scientist

All Animals Need Food to Survive

Read the text. Then, **record** other questions you have about how energy transfers from a food source to a living organism.

Quick Code: us5281s

All Animals Need Food to Survive

All animals need food to survive. Do plants need food to survive? Where does the food come from? Does it come from a plant, an animal, or somewhere else? Just what is food anyway, and why do we need it? Is food always something that's living, such as a fish or a carrot? Do we need things other than food to live? These are some of the questions you'll be able to answer in this concept.

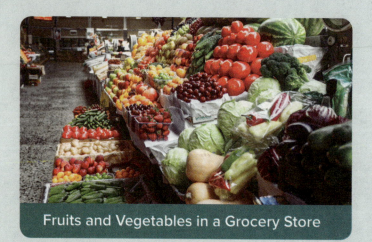

Fruits and Vegetables in a Grocery Store

SEP Asking Questions and Defining Problems
CCC Patterns

My Questions

2.2 | Wonder How does matter flow through an ecosystem?

Activity 4
Observe Like a Scientist

Decay

Look at the images. Then, **answer** the question.

Quick Code: us5282s

Decomposing Seagull

Rotting Logs

Leaf Mold

What do these images have in common?

_____

_____

_____

CCC Patterns

Activity 5

Evaluate Like a Scientist

Quick Code: us5283s

What Do You Already Know About Matter Flow in Ecosystems?

What Do They Eat?

Match each food or foods to the animal that eats it.

Animals

Food

Concept 2.2: Matter Flow in Ecosystems | 57

2.2 | Wonder — How does matter flow through an ecosystem?

Why Eat Plants or Animals?

Read the question. Then, **write** your answer in the space provided.

Why do animals eat plants or other animals?

_____
_____
_____
_____
_____
_____
_____
_____
_____
_____
_____
_____

Ecosystems

Read the questions. Then, **write** your answers in the space provided.

What is an ecosystem?

_____

_____

_____

_____

Write some examples of ecosystems.

_____

_____

_____

_____

_____

_____

2.2 | Learn How does matter flow through an ecosystem?

 Activity 6
Investigate Like a Scientist

Quick Code: us5284s

Hands-On Investigation: What's Happening to the Bread?

In this investigation, you will **observe** the process of decomposition by mold. You will **investigate** the environmental conditions under which mold decomposes bread most rapidly. You will **formulate** a hypothesis and **test** your hypothesis by observing the growth of mold on bread over the course of a week. You will **set up** the activity today and complete the activity later in this concept.

Make a Prediction

Develop a claim about what you think will happen to the bread.

Which conditions do you think will encourage more mold growth than the control piece of bread?

_____

_____

Which conditions do you think will encourage less mold growth than the control piece of bread?

_____

_____

SEP Analyzing and Interpreting Data
SEP Planning and Carrying Out Investigations

What materials do you need? (per group)

- Fresh bread (without preservatives)
- Plastic zipper bags
- Various environmental conditions such as light, darkness, heat, cold, humidity, etc.

What Will You Do?

Follow these steps to test your claim.

1. Discuss different environmental conditions under which mold can grow on bread, such as light, darkness, heat, cold, humidity, and acidity. Choose 4–6 different conditions to test. Include a control.

2. Create a hypothesis about whether each condition would encourage more mold growth or less than the control.

3. Place a slice of bread in each of the environmental conditions you identified. Seal each slice of bread in a plastic bag.

4. Create a table to record your data. Include the day number, a description of the bread's appearance, and a sketch. Make daily observations of the bread and record your observations in the table.

Concept 2.2: Matter Flow in Ecosystems | 61

2.2 | Learn — How does matter flow through an ecosystem?

Day	Appearance of Bread Description	Sketch
1		
2		
3		
4		
5		

After collecting your final observations, analyze your data. Discuss the different conditions your classmates investigated and draw conclusions.

Think About the Activity

What conditions were most favorable to mold growth and its decomposition of bread?

_____
_____
_____
_____
_____

How do your results compare with your hypothesis?

_____
_____
_____
_____

2.2 | Learn — How does matter flow through an ecosystem?

Many types of bacteria are also decomposers. Based on your results, can you predict what types of environments would be most favorable to bacterial decomposition?

_____
_____
_____

How Do Food Webs Show the Relationships between Living Things?

Activity 7
Analyze Like a Scientist

Quick Code: us5285s

Food Is Energy

Read the text with a partner, and then together **determine** the main idea for each paragraph. **Underline** the main idea for each paragraph.

Food Is Energy

How do you get the energy you need to think, breathe, move, or anything else? Some activities such as hard work or exercise require a lot of energy. But even when you sleep, your body still uses some energy. Food and the oxygen we breathe provide the energy we need to do any kind of activity.

Runner

CCC Energy and Matter

Concept 2.2: Matter Flow in Ecosystems | 65

Food Is Energy *cont'd*

Leaf in Sun

Grizzly Bear Eating

All living things need energy to live, grow, and do other work. The source of energy for all organisms on Earth is the **sun**. Plants **absorb** sunlight through their leaves and use the sun's energy to make their own food. Sunlight provides the energy for plants to convert water and one of the gases in air into sugar, or food. This process is fundamental to life on Earth.

Animals, like humans, cannot make their own food. Instead, they get their energy from the **environment** in which they live. Some animals eat plants as food. Some eat other animals that eat plants. Some eat both plants and animals! In this way, energy produced from the sun passes through all life on Earth.

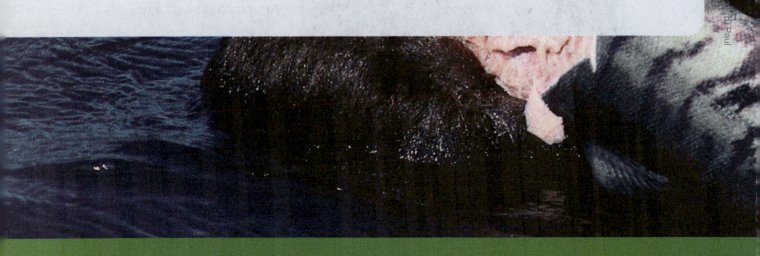

Activity 8

Think Like a Scientist

Food Webs in the Neighborhood

In this investigation, you will **observe** a habitat in your neighborhood and **identify** food webs in that environment. You will **identify** feeding relationships between organisms and **make a model** that shows those relationships.

Quick Code: us5286s

What materials do you need? (per group)
- Hand lens
- Colored pencils
- Camera

SEP Developing and Using Models

2.2 | Learn How does matter flow through an ecosystem?

What Will You Do?

1. With your group, generate ideas about organisms that you will need to look for to produce a food web of the ecosystem. Consider what types of plants or animals you expect to find.

2. Formulate questions to guide your investigation. Record your questions and refer to them as you complete the activity.

3. Explore an outdoor area. Move slowly and do not disturb the environment. Note the different types of organisms that live there. Pay particular attention to energy relationships in the environment. Record the relationships in your notebook and with your camera.

4. In class, arrange the organisms you observed in a food web. Print out pictures or copy sketches from your observations to form the notes of the web. Document on your food web any feeding activities you observed directly. Fill in missing relationships on your food web by researching the predators and prey of the organisms you have identified.

Food Web

Organism	Feeding Activity Observations	Sketch

2.2 | Learn   How does matter flow through an ecosystem?

Think About the Activity

What organisms did you place in your food web, and how are they related to one another?

_____
_____
_____
_____
_____
_____
_____

What types of plants, both living and dead, did you observe? What can you infer about the needs of these organisms?

_____
_____
_____
_____
_____
_____
_____

Choose an organism, and **remove** it from your food web. What do you think would happen to the ecosystem when that organism is removed?

_____
_____
_____
_____
_____
_____
_____
_____
_____
_____
_____
_____

2.2 | Learn How does matter flow through an ecosystem?

Activity 9
Observe Like a Scientist

Food Chains

Quick Code: us5287s

Watch the video. As you watch, **look** for an example of a food chain.

Food Chains

Talk Together

Now, talk together about the role of each organism in a food chain.

CCC Energy and Matter

Activity 10

Analyze Like a Scientist

Energy Flow

Quick Code: us5288s

Read the text. **Highlight** evidence you can use to support your answer to what would happen if an organism in your local food web was removed. **Record** the evidence in the space provided.

Energy Flow

All organisms need energy. Organisms that don't capture energy directly from the sun need other organisms to obtain energy. Food chains show how energy passes from one organism to another in an **ecosystem**. They show the food, or energy, relationships among organisms within specific ecosystems.

Grass makes its own food using energy from sunlight. A mouse eats the grass to get energy. A snake then eats the mouse, and a hawk then eats the snake. The energy from the sun passes to the grass, then to the mouse and snake, and finally to the hawk. Unlike grass, mice cannot make their own food from sunlight. The following **food chain** shows the relationship among these organisms.

grass ⟶ mouse ⟶ snake ⟶ hawk

Food Chain

In this food chain, the hawk and the snake are predators. They hunt other animals as **prey**. The snake and the mouse are prey. They are hunted by other animals as food. Both predators and prey pass food and energy through the food chain.

Concept 2.2: Matter Flow in Ecosystems

Rattlesnake

My Evidence

CCC Energy and Matter

Activity 11
Evaluate Like a Scientist

Food Chain

Write the names of the organisms in the correct boxes to make a food chain.

Quick Code: us5289s

| bird | grass | grasshopper | owl | snake |

☐ → ☐ → ☐ → ☐ → ☐

How would you add a grass-eating beetle that the bird eats to this model?

_____
_____
_____
_____
_____

SEP Developing and Using Models

Activity 12
Analyze Like a Scientist

Food Webs

Quick Code: us5290s

Read the text. **Think** about the organisms you observed or read about in this lesson. Then, **write** the names of organisms you observed or read about in the correct column of the T-Chart.

Food Webs

Sometimes, we draw concept webs or main idea webs to show the relationships among different bits of information. We can also show the interactions among organisms for food in a similar way. Think about the different foods you eat. Imagine those foods are connected to you by lines in a web. All living things, including you, **interact** in **food webs**. We can draw these webs to show how organisms are connected within ecosystems.

Hawks are part of a forest food web.

Food webs are made up of food chains. Food chains show the relationship of food and energy that passes from one organism to another. All food chains begin with an energy source, like the sun. The sun provides energy for the next link in the chain, the **producer**. Plants are producers. Plants provide food for a series of consumers, which may eat only plants or may eat both plants and animals. The ways in which many food chains intersect within an ecosystem form a food web.

Producers	Predators

Revise a model of how the hawk interacts with the environment. You may add organisms to the model. Use vocabulary from the text. You can use words, images, and/or symbols.

My Model

SEP Developing and Using Models

Concept 2.2: Matter Flow in Ecosystems

2.2 | Learn How does matter flow through an ecosystem?

Activity 13

Observe Like a Scientist

Forest Food Web

Quick Code: us5291s

Look at the image. Then, **answer** the questions that follow.

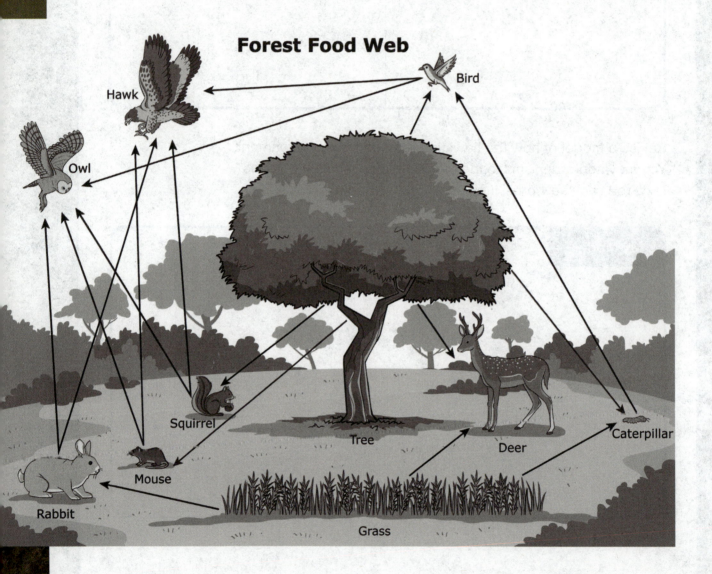

CCC Energy and Matter

What would happen to the rabbit if all the grass were removed from the area?

___

What would happen to the hawk if all the grass were removed from the area?

___

How does energy travel from the grass to the hawk?

___

Concept 2.2: Matter Flow in Ecosystems

2.2 | Learn — How does matter flow through an ecosystem?

Activity 14

Evaluate Like a Scientist

Food Webs and Competition

Quick Code: us5292s

Answer the three questions to help you organize your ideas about food webs.

How do food webs model competition among organisms in an ecosystem?

___

___

___

___

How does a food web represent a system for the transfer of energy?

___

___

___

___

___

___

___

Why is a food web a better choice to use to show interactions among organisms than food chains?

_____

_____

_____

_____

Now, **draw** a diagram of your own food web for an ecosystem of your choosing. Be sure to include at least five different organisms in your food web.

SEP Developing and Using Models

Why Are Decomposers Important in an Ecosystem?

Activity 15
Analyze Like a Scientist

Quick Code: us5293s

What Are Decomposers?

Read the text. **Underline** any characteristics of a decomposer.

What Are Decomposers?

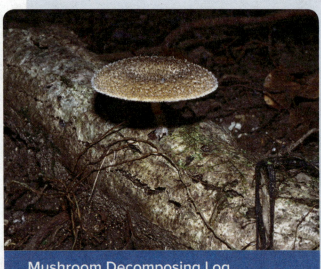

Mushroom Decomposing Log

Have you ever seen mold growing on a piece of bread or mushrooms growing in soil? If you have, then you have seen decomposition in action. Scavengers, such as vultures, hyenas, crabs, cockroaches, and houseflies, begin the process of decomposition. They break food down into smaller pieces. Then, **decomposers**, including snails,

CCC Energy and Matter

slugs, earthworms, fungi, and bacteria, complete the process and eat the remains of dead plants and animals.

Decomposers play a vital part in the environment. They help break down dead plants and animals into nutrients that can be returned to the ecosystem. Plants use the nutrients, and the **cycle** continues from producers to consumers to decomposers and back to producers again. Recall that this complicated relationship between different organisms in an ecosystem is called a food web.

When you are finished using something like a food wrapper, or a piece of paper, you might throw it into a trash can. From there, the trash is taken to a landfill with all the other trash. Humans produce a lot of waste, so landfills have to take up more and more space. One way that people reduce this waste is by recycling. When you **recycle** something, it gets used to make new products instead of going into a landfill.

A similar thing happens in natural environments. Without decomposers, dead things would build up, just like the trash in landfills. Decomposition is nature's recycling factory. Living things contain nutrients, the chemicals that all organisms need to survive and grow. The world has a limited amount of nutrients that can be used by living things. When organisms die, decomposition releases these nutrients back into the environment, so they can be used again. For example, decomposed animal and plant remains become part of the soil, which is used by plants. Decomposition can take place underwater, too. Underwater, decomposed living things fall to the bottom of the body of water and provide nutrients to underwater plants.

Concept 2.2: Matter Flow in Ecosystems | 83

2.2 | Learn How does matter flow through an ecosystem?

Activity 16

Observe Like a Scientist

Decomposition

Complete the interactive to learn about decomposition. **Observe** the different steps involved in tree decomposition. As you work, **sketch** and **label** each step.

Quick Code: us5294s

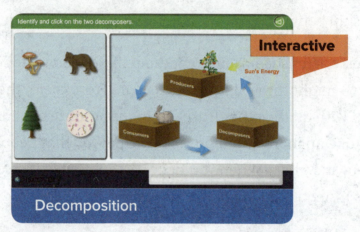

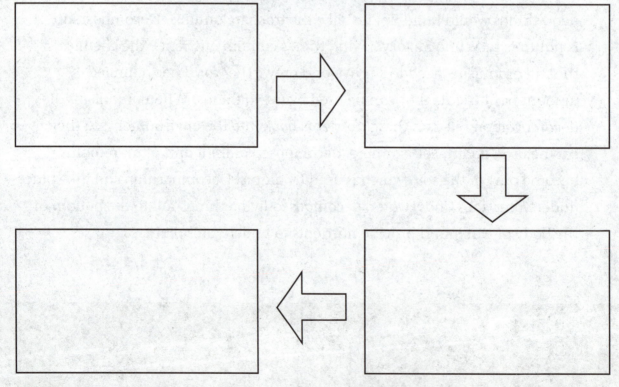

CCC Energy and Matter

Now, **use** your observations to **answer** the questions.
What are decomposers? Name two examples of decomposers.

_____
_____
_____
_____
_____

Explain the process of decomposition.

_____
_____
_____
_____
_____
_____

What would happen if there were no decomposers?

_____
_____
_____
_____
_____

2.2 | Learn How does matter flow through an ecosystem?

Activity 17
Observe Like a Scientist

Decomposers

Quick Code: us5295s

Watch the video. **Look** for evidence of how decomposers help move energy in an ecosystem.

Decomposers

Talk Together

Now, talk together about how decomposers help move energy through an ecosystem. How can you show this on your ecosystem models?

How Do Living Things Depend on Nonliving Things to Survive in an Ecosystem?

Activity 18
Analyze Like a Scientist

Quick Code: us5296s

Living Things Depend on Nonliving Things

Make a claim about whether living things depend on nonliving things. **Read** the text. **Highlight** evidence that supports or disputes your claim. **Write** the evidence in the C-E-R table that follows.

Living Things Depend on Nonliving Things

Think about the nonliving things you depend on each day. You need to breathe air. You need to drink water. You also use water to wash. Our buildings and roads are built on solid rock. Other living things depend on nonliving parts of their environment, too.

Deer Eating Leaves

SEP Engaging in Argument from Evidence

Concept 2.2: Matter Flow in Ecosystems | 87

Living Things Depend on Nonliving Things *cont'd*

An ecosystem includes both living and nonliving things that interact together in an environment. Animals also need space to live and find food. They also need **shelter** within their space, so they have a safe place to hide, sleep, or raise their young. Plants need enough space for their roots and leaves. When all of their needs are met, organisms have a chance to live and reproduce.

In addition to nonliving things such as water, air, and space, all organisms need nutrients and energy. Nutrients are nonliving substances such as minerals, fats, proteins, or carbohydrates that living things need to survive. Animals get their nutrients by eating plants. Plants get their nutrients from the soil. Plants absorb these nutrients through their roots as they grow. When a plant dies and is broken down by decomposers, those nutrients are released back into the soil. When a new plant grows in that soil, there are plenty of nutrients for it. In this way, decomposers return nutrients to the soil.

Another essential nonliving thing plants need is energy from sunlight. Animals don't depend directly on sunlight, but they rely on it indirectly when they eat plants. Plants use this energy from sunlight, water and nutrients absorbed from the roots, and a gas from the air to make food in their leaves. This food serves as food for the plant but also food for animals that might eat the plant.

My claim

Evidence I have found:

Reasoning: My claim is true because:

Concept 2.2: Matter Flow in Ecosystems | 89

2.2 | Learn How does matter flow through an ecosystem?

Activity 19
Observe Like a Scientist

Quick Code: us5297s

Nonliving Needs

Watch the videos. **Look** for evidence for your claim. **Write** the evidence in your Claim-Evidence-Reasoning table from Activity 18.

Need for Food and Water

Need for Air

Need for Living Space

SEP Engaging in Argument from Evidence

Now, **use** the evidence in your C-E-R table to **write** a paragraph that shows your claim is supported by evidence.

Concept 2.2: Matter Flow in Ecosystems

2.2 | Learn How does matter flow through an ecosystem?

Activity 20
Evaluate Like a Scientist

Quick Code: us5298s

Comparing Needs

Plants and animals have some of the same needs and some different needs. **Write** each need in the Venn Diagram to show if only plants need it, if both need it, or if only animals need it.

| water | food | sunlight | carbon dioxide |
| oxygen | space | shelter | |

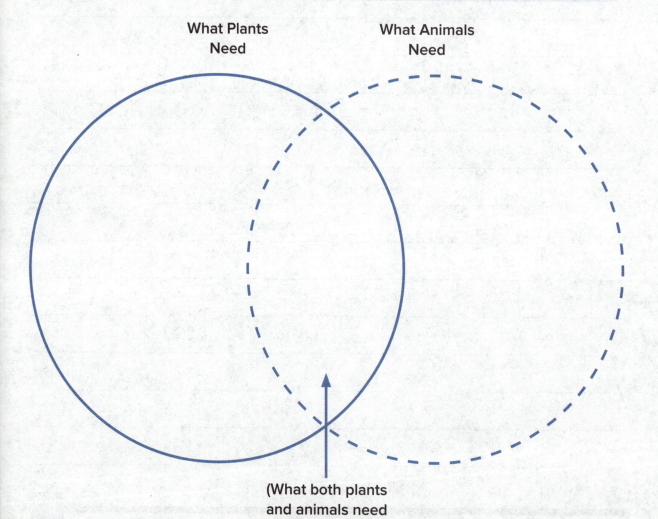

What Plants Need

What Animals Need

(What both plants and animals need goes in the middle.)

Concept 2.2: Matter Flow in Ecosystems

2.2 | Share How does matter flow through an ecosystem?

Activity 21
Record Evidence Like a Scientist

Quick Code: us5299s

How Hawks Find Food

Now that you have learned about plant needs, **look** again at the image How Hawks Find Food. You first saw this in Wonder.

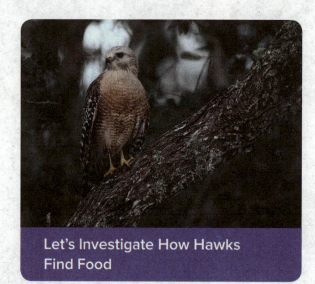

Let's Investigate How Hawks Find Food

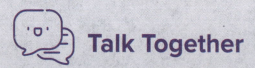

How can you describe how hawks interact with their environment now? How is your explanation different from before?

SEP Constructing Explanations and Designing Solutions

Look at the Can You Explain? question. You first read this question at the beginning of the lesson.

> **Can You Explain?**
>
> How does matter flow through an ecosystem?

Now, you will use your new ideas about How Hawks Find Food to answer a question.

1. **Choose** a question. You can use the Can You Explain? question or one of your own. You can also use one of the questions that you wrote at the beginning of the lesson.

   My Question

2. Then, **use** the C-E-R table on the next page to answer the question.

Concept 2.2: Matter Flow in Ecosystems | 95

2.2 | Share How does matter flow through an ecosystem?

Work with a partner to **write** a claim and **gather** evidence to the Can You Explain? question. Brainstorm what science words you want to use in your reasoning. Use the table below to **record** your responses.

My claim (the answer to the question)
Evidence I have found (record all the evidence you gathered from video, reading, interactives, and Hands-On Investigations)

Reasoning: My claim is true because:

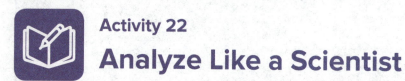

STEM in Action

Quick Code: us5300s

Activity 22
Analyze Like a Scientist

A Backyard for Bees, Birds, and Butterflies

Read the text. **Watch** the video. **Look** for information about bird habitats.

A Backyard for Bees, Birds, and Butterflies

Video

Backyard Habitat: City Birds

When we think about animals and their needs, we often think of our pets. We might also think of animals that live in zoos. Or maybe we think of animals that live in wild areas. We may not think of the many animals that live in cities. These animals have to meet their needs in an environment that is constructed for human residents.

SEP Obtaining, Evaluating, and Communicating Information

Some people who live in cities want to help animals and plants meet their needs. For example, people love to watch colorful birds and hear their songs in the summer. By creating a backyard that meets the needs of birds, they might be able to see more of these fascinating creatures.

Landscape architects design backyard habitats. The habitat can be for birds or other animals, such as bees and butterflies. They use computer-aided design (CAD) software to design a plan. Then, they show the plan to the homeowners. After the plan is approved, they get to work. The plants and structures they include depend on what kinds of wildlife the homeowner wants to attract. For example, they might add milkweed, which is an important plant for monarch butterflies. They might add native flowering plants to increase honeybee populations. They might add bat boxes if the homeowner wants to attract bats, which eat mosquitoes.

 Talk Together

Now, talk together about backyard habitats. What animals have you seen in your backyard?

Winter Challenges

Robert and Tanika have a birdbath in their backyard. It provides water for songbirds. However, in the winter, the water freezes. How can Robert and Tanika use technology to solve their problem?

Write a description of a device that they could use to solve the problem. In your description, **include** how your device gets the energy it needs to work. Then, **create** a model (in the form of a labeled drawing) of your device.

Birdbath

Description:

_____
_____
_____
_____
_____
_____

Diagram:

Concept 2.2: Matter Flow in Ecosystems

2.2 | Share — How does matter flow through an ecosystem?

Activity 23
Evaluate Like a Scientist

Quick Code: us5301s

Review: Matter Flow in Ecosystems

Think about what you have read and seen. What have you learned?

Write down some key ideas you have learned. **Review** your notes with a partner. Your teacher may also have you take a practice test.

SEP Obtaining, Evaluating, and Communicating Information

Talk Together

Think about what you saw in Get Started. Use your new ideas about matter flow in ecosystems to discuss how organisms are connected.

CONCEPT
2.3

Energy Flow in Ecosystems

Student Objectives

By the end of this lesson:

☐ I can develop models that demonstrate that energy from the sun is transformed by living things into energy in animals' food.

☐ I can construct an explanation based on mathematical and computational thinking and logical reasoning for the small amount of energy transferred between organisms in a food web.

Key Vocabulary

☐ chemical energy
☐ energy
☐ energy pyramid
☐ energy transfer
☐ light
☐ mechanical energy

☐ nuclear energy
☐ nucleus
☐ radiant energy
☐ radiation
☐ sound
☐ substance

☐ terrarium

Quick Code: us5303s

Concept 2.3: Energy Flow in Ecosystems | 105

Activity 1
Can You Explain?

How can the energy animals use for body repair, growth, motion, and body warmth be traced back to the sun?

_____
_____
_____
_____
_____
_____
_____
_____
_____
_____

Quick Code: us5304s

Concept 2.3: Energy Flow in Ecosystems | 107

2.3 | Wonder

How can the energy animals use for body repair, growth, motion, and body warmth be traced back to the sun?

Activity 2
Ask Questions Like a Scientist

The Sun and Plants

Quick Code: us5305s

You have heard a teacher or friend tell you that the sun is important to life on Earth. What do you wonder about the sun? **Write** down what you wonder about the sun.

Wonder Statements

SEP Asking Questions and Defining Problems
CCC Energy and Matter

Watch the video. Then, **complete** the activity that follows.

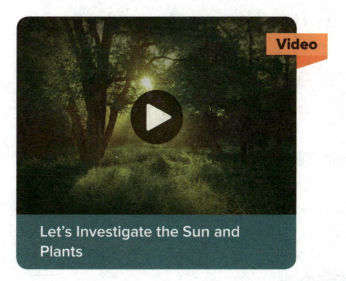

Let's Investigate the Sun and Plants

Write the following claim on your graphic organizer: "All organisms on Earth depend on the sun." As you work through the concept, **collect** evidence to add to your graphic organizer.

Claim

Evidence

Reasoning

Concept 2.3: Energy Flow in Ecosystems | 109

2.3 | Wonder
How can the energy animals use for body repair, growth, motion, and body warmth be traced back to the sun?

Activity 3
Evaluate Like a Scientist

What Do You Already Know About Energy Flow in Ecosystems?

Quick Code: us5306s

Where Do They Get Energy?

Look at the organisms on the left. **Write** the names of the organisms in the blanks to show which one eats which. Which organism goes at the top of the list? Write the organism it eats underneath. Continue doing this until all the organisms are listed.

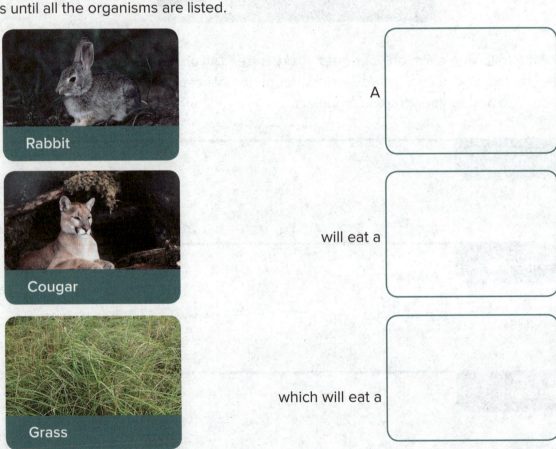

A _____

_____ will eat a

_____ which will eat a

_____ which will eat _____

Food Webs

Look at the image and **describe** which organisms eat other organisms.

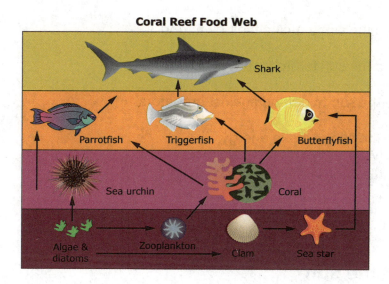

SEP Constructing Explanations and Designing Solutions
CCC Energy and Matter

Concept 2.3: Energy Flow in Ecosystems

2.3 | Wonder
How can the energy animals use for body repair, growth, motion, and body warmth be traced back to the sun?

Plant Life

Check all the statements that correctly describe plants and their relationship with the sun.

- ☐ Plants need the sun to make their own food.
- ☐ Plants need the sun to produce water.
- ☐ Plants use the sun to defend against predators.
- ☐ Plants use the sun to make energy.
- ☐ Plants will die if there is not enough sunlight.
- ☐ Plants will die if there is too much sunlight.

Life without the Sun

What would happen to life on Earth if there was no sunlight?

_____

_____

_____

_____

Concept 2.3: Energy Flow in Ecosystems

2.3 | Learn
How can the energy animals use for body repair, growth, motion, and body warmth be traced back to the sun?

How Does Energy from the Sun Flow through an Ecosystem?

Activity 4
Evaluate Like a Scientist

Quick Code: us5307s

Forms of Energy

Each situation below describes an example of potential or kinetic energy. **Circle** the correct form of energy each example represents.

A battery is hooked up to a lighted light bulb.	Potential	Kinetic
A car is coasting down a long hill.	Potential	Kinetic
A battery is sitting on a shelf in a store.	Potential	Kinetic
The sun is shining during the day.	Potential	Kinetic
A car is parked near the top of a hill.	Potential	Kinetic
The ground is cooling off at night.	Potential	Kinetic

Activity 5

Think Like a Scientist

Energy Flow

Quick Code: us5308s

In this activity, you will model the flow of energy through a food web.

What materials do you need? (per group)

- Index cards
- Food web
- Marbles, $\frac{5}{8}$ in.

What Will You Do?

1. Use your marbles as energy content.
2. Play a game of predator/prey, where you capture or lose your energy (marbles).

Talk Together

What fraction of your energy did you pass on to a predator? Why is the sun necessary for food webs to maintain themselves?

CCC Energy and Matter

Activity 6
Analyze Like a Scientist

Quick Code: us5309s

Forms of Energy

Read the text. As you read, **underline** the different types of energy you read about along with their definitions. Then, **complete** the activity that follows.

Forms of Energy

If you look around, you can probably see many things that are happening because of **energy**. Energy is all around us all the time. There are many types of energy. For example, **radiant energy** comes from the sun as **light** energy and other types of **radiation**. Thermal energy heats our homes. A moving car has motion energy. **Sound** is actually vibrations caused by waves of energy around us. Electrical energy comes from the movement of tiny charged particles either through a wire or, in the case of lightning, through air.

There are some other types of energy you might not think about very often. For example, **chemical energy** is the energy stored in chemical bonds between atoms and

Hummingbird

CCC Energy and Matter

molecules. This is the type of energy that is contained in gasoline and food. **Nuclear energy** is the energy that holds the **nucleus** of an atom together. **Mechanical energy** is the energy of objects due to motion, such as moving fan blades or a soaring bird. Whether we think about it or not, each type of energy plays an important role in our daily lives.

Gasoline provides energy for a car to move.

But just what is energy? Scientists define energy as the ability to do work or to cause some kind of change. This often means that energy can make things move, such as when a motor makes a fan spin. Or, energy can cause something to change, such as when a fire cooks meat. Energy comes from something, called an energy source (like gasoline), and then goes to something else, called an energy receiver (like a car). In this way, energy is transferred from one place or object to another place or object. So, when gasoline provides energy for a car to move, we can draw a diagram like this:

| Gasoline | → | Car |

Concept 2.3: Energy Flow in Ecosystems

Forms of Energy *cont'd*

This is called an energy chain because it shows the transfer of energy from the energy source (the gasoline) to the energy receiver (the car) and it resembles the links of a chain. Energy chains can have many parts to them, like this:

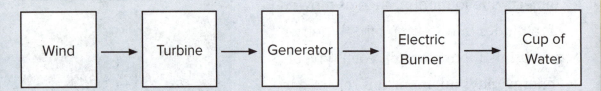

This energy chain shows the flow of energy from moving air (wind) to the blades of a wind turbine, which spins a generator. The generator produces electricity, which heats an electric burner that in turn heats water for hot chocolate.

Food chains are actually energy chains. This food chain:

is an energy chain. The energy that was in the grass is transferred to the body of a rabbit when the rabbit eats the grass. The energy that is in the rabbit's body then is transferred on to the hawk when the hawk eats the rabbit.

Now, **complete** the table with the information for each type of energy.

	Definition	Example	Relevant Terms or Vocabulary
Energy			
Radiant Energy			
Electrical Energy			
Chemical Energy			
Nuclear Energy			
Mechanical Energy			

Concept 2.3: Energy Flow in Ecosystems

2.3 | Learn
How can the energy animals use for body repair, growth, motion, and body warmth be traced back to the sun?

Activity 7
Observe Like a Scientist

Energy in Systems

Quick Code: us5310s

Complete the interactive. Then, **draw** your own energy chains: one for living systems and one for nonliving systems.

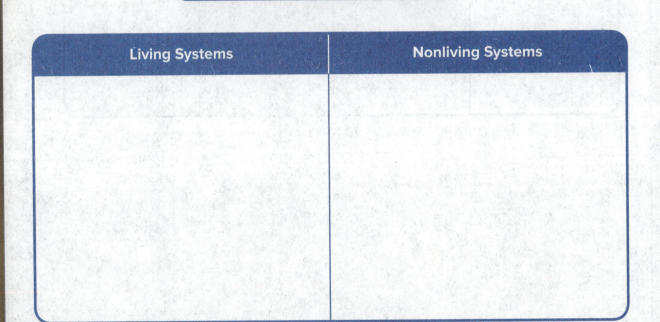

Energy in Systems

Living Systems	Nonliving Systems

CCC Energy and Matter

Activity 8

Evaluate Like a Scientist

Quick Code:
us5311s

Bread: An Energy Story

Energy flows from the sun to a mouse eating bread. Decide the main kind of energy in each picture and the system it represents. **Draw** a line to match each picture to the energy form that describes it.

Sun Shining on Wheat Fields

Wheat Growing

Harvester Working

Baking the Bread

Eating the Bread

Light Energy

Heat Energy

Chemical Energy

CCC Energy and Matter

Concept 2.3: Energy Flow in Ecosystems | 121

2.3 | Learn
How can the energy animals use for body repair, growth, motion, and body warmth be traced back to the sun?

Create a model showing how energy moves through the process of making bread. Your model can contain words, images, and/or symbols.

My Model

Activity 9
Analyze Like a Scientist

Energy Pyramid

Quick Code: us5312s

Read the text. As you read, **draw** arrows on the energy pyramid to represent the flow of energy in a way that demonstrates that only 10 percent gets passed on to each level.

Energy Pyramid

Energy flows through a food chain, but only about 10 percent of it is passed to the next level. Each organism uses 90 percent of its energy to live and to grow. This **energy transfer** can be represented by an **energy pyramid**, with a lot of energy at the bottom and a much smaller amount on top.

The small amount of energy transferred between organisms in a food web means that a constant input of the sun's energy is needed. This is why the sun is necessary not only for plants, but for all life on Earth.

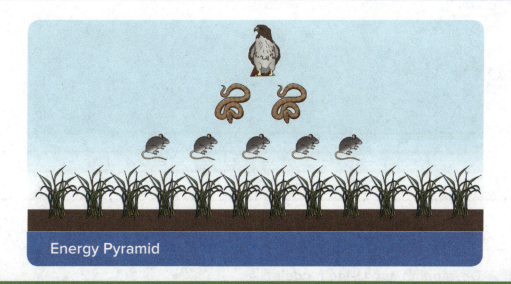

Energy Pyramid

Concept 2.3: Energy Flow in Ecosystems

2.3 | Learn
How can the energy animals use for body repair, growth, motion, and body warmth be traced back to the sun?

Activity 10

Evaluate Like a Scientist

Quick Code: us5313s

A Pyramid of Organisms

Look at the following image of an Energy Pyramid. Use this image to **draw** your own pie chart of the number of organisms found in the energy pyramid. Be sure to label each part of your chart.

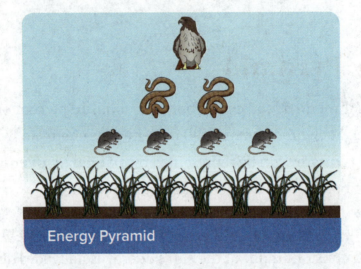

Energy Pyramid

SEP Developing and Using Models

How Do Animals Obtain the Energy They Need to Live?

 Activity 11
Think Like a Scientist

Quick Code: us5314s

Modeling the Flow of Energy and Matter in an Ecosystem

In this activity, you will create a model of an ecosystem to display how matter and energy flow within that ecosystem.

What materials do you need? (per group)

- Index cards
- String
- Scissors
- Connecting cubes
- Pencils

SEP Developing and Using Models
CCC Energy and Matter

Concept 2.3: Energy Flow in Ecosystems | 125

2.3 | Learn

How can the energy animals use for body repair, growth, motion, and body warmth be traced back to the sun?

What Will You Do?

1. With your group, label the notecards: sun, 10 producers, 3 consumers, and 2 decomposers in the ecosystem you have been assigned.

2. Create a food web by connecting the appropriate notecards with string.

3. Present your model to the class.

4. Use the cubes to represent matter in soil and air.

5. Place the cubes on your cards to represent what each organism in your ecosystem consumes.

Think About the Activity

Describe the flow of energy in your model ecosystem. How is it different than the flow of energy?

_____
_____
_____
_____

Describe the flow of matter in your model ecosystem.

_____
_____
_____
_____
_____
_____

What effect would the addition of more consumers have on the ecosystem?

_____
_____
_____
_____
_____
_____
_____

Describe the role of decomposers and **explain** why they are important.

_____
_____
_____
_____

What would happen if the producers were removed from the ecosystem?

_____
_____
_____

2.3 | Learn

How can the energy animals use for body repair, growth, motion, and body warmth be traced back to the sun?

Activity 12
Evaluate Like a Scientist

Quick Code: us5315s

Food Web

Look at the food web. **Write** numbers 1 through 5 to list the organisms from the highest amount of energy (1) to the lowest (5).

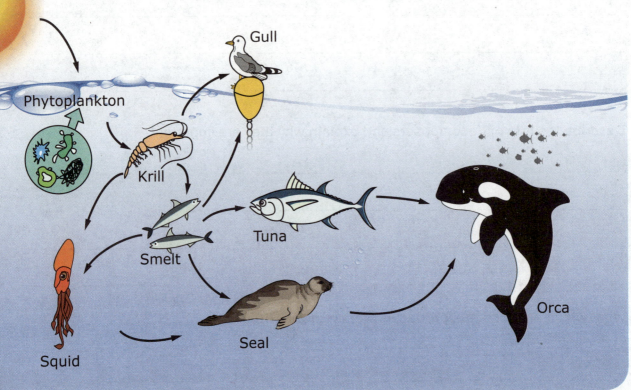

SEP Developing and Using Models

Concept 2.3: Energy Flow in Ecosystems

2.3 | Share
How can the energy animals use for body repair, growth, motion, and body warmth be traced back to the sun?

Activity 13
Record Evidence Like a Scientist

The Sun and Plants

Now that you have learned about energy flow in ecosystems, **watch** the video about sun and plants again. You first saw this in Wonder.

Quick Code: us5316s

Let's Investigate the Sun and Plants

Talk Together

How can you describe the Sun and Plants video now? How is your explanation different from before?

SEP Constructing Explanations and Designing Solutions

Look at the Can You Explain? question. You first read this question at the beginning of the lesson.

> **Can You Explain?**
>
> How can the energy animals use for body repair, growth, motion, and body warmth be traced back to the sun?

Now, you will use your new ideas about the sun and plants to answer a question.

1. **Choose** a question. You can use the Can you Explain? question or one of your own. You can also use one of the questions that you wrote at the beginning of the lesson.

My Question

Concept 2.3: Energy Flow in Ecosystems | 131

2.3 | Share
How can the energy animals use for body repair, growth, motion, and body warmth be traced back to the sun?

2. **Revisit** your claim about how all organisms on Earth depend on the sun from the graphic organizer in Wonder.

3. Next, **list** your evidence and **explain** your reasoning. Reasoning ties together the claim and the evidence. Reasoning shows how or why the data count as evidence to support the claim.

Topic: _____

Evidence	Reasoning

4. You can use this information to write your scientific explanation. First, **write** your claim. Then, complete your explanation using your evidence and reasoning.

Now, **write** your scientific explanation.

_____
_____
_____
_____
_____
_____
_____
_____
_____
_____

STEM in Action

Activity 14
Analyze Like a Scientist

Quick Code: us5317s

Food for Thought

Read the text. Then, **complete** the activity that follows.

Food for Thought

You have learned that humans are consumers who tend to eat both plants and animals. However, think about what you eat every day. Do you always eat the same food? Probably not. Most nonhuman consumers do not eat the same thing every day either. They eat a variety of things, just like we do. Producers and consumers are part of many food chains, which is why it is more realistic to model energy flow using a food web.

Food scientists study the basic elements of food webs using chemistry and biology. They help us understand what foods are made of and which foods are healthy to eat. Now that you know more about how we are connected to the plants and animals we eat, let's learn more about the importance of studying food science.

SEP Analyzing and Interpreting Data

Studying food and nutrition also gives us information about the calories in our food. You have probably heard the term calorie. Calories are the way we measure the amount of energy in various foods. Knowing how many calories we eat is important. We need a certain number of calories to carry out our daily activities. If we eat too many calories, our bodies will store those calories for use later, as fat. Food technologists measure the energy, or the calories, in foods we eat. They use a special instrument, as you will see.

Food and Energy

As the number of humans increases, we need to learn even more about growing enough food to feed everyone. Scientists are developing ways to modify some of the main crops we eat.

Rice is one of the main crops grown to feed people around the world, especially in Asia. As the human population increases, the land available to grow rice crops decreases. Rice, like most crops, needs a lot of space to grow. However, much of the land available for farming in Asia has many hills.

Thinking Like a Scientist

Look at these two images of rice fields. Imagine you are a rice farmer, and you have two plots of land. The two plots are the same size. One of the plots of land has hills and very little space near the ground. You decide to build terraces, or ledges, along the sides of the hills. At the end of the growing season, you discover that one of the plots of land produced almost two times more rice! How is this possible? Which plot of land produced more rice? **Explain** your answer.

Flat Rice Field

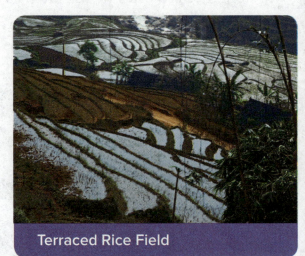

Terraced Rice Field

Activity 15

Evaluate Like a Scientist

Review: Energy Flow in Ecosystems

Think about what you have read and seen in this lesson. **Write** down some core ideas you have learned. **Review** your notes with a partner. Your teacher may also have you take a practice test.

Quick Code: us5318s

_____

_____

_____

_____

_____

_____

_____

Talk Together

Think about what you saw in Get Started. Use your new ideas to discuss how organisms are connected in ecosystems.

SEP Obtaining, Evaluating, and Communicating Information

Unit Project

Solve Problems Like a Scientist

Unit Project: Build a Mini-Ecosystem

In this project, you will use what you have learned about ecosystems and energy flow to build your own mini-ecosystem. **Read** the text and **watch** the video about terrariums. Think about what you need to build a model ecosystem. Then, **complete** the activities that follow.

Quick Code: us5320s

Mini-Ecosystems

SEP Developing and Using Models
CCC Energy and Matter

Build a Mini-Ecosystem

Many ecosystems exist around the world and around us. Many different interactions take place among the different variables in an ecosystem. These variables consist of living and nonliving factors. Each ecosystem is unique. The types of biotic and abiotic factors, ways in which they interact and other things, contribute to this uniqueness. Think about an ecosystem, such as a tropical rain forest or a desert. What types of living and nonliving factors are present? How do these factors interact with one another in an ecosystem?

Making a Terrarium

In this project, you will construct your own mini-ecosystem to build upon what you already know about the interactions between organisms and energy flow within an ecosystem. You will use a container to house your ecosystem. After your teacher explains how you should construct your mini-ecosystem, you will create a diagram. This diagram will show energy relationships in your ecosystem.

Unit Project

Ecosystem Building

Which statement accurately describes the most important purpose of creating your mini-ecosystem? **Circle** the correct answer.

   A. To model what happens in a real-life ecosystem
   B. To show how only abiotic factors interact with each other
   C. To indicate which items in the ecosystem are made of matter
   D. To illustrate how biotic factors are initial sources of energy

How will you design your own model ecosystem?

_____
_____
_____
_____
_____
_____

What materials will you need for your design?

_____
_____
_____
_____
_____
_____

Draw a plan of your mini-ecosystem. **Label** all the parts of your design.

Then, **build** an ecosystem according to your plan.

Unit 2: From Matter to Organisms | 141

Unit Project

Energy Flow Diagram

After building your ecosystem, **think** about how energy flows through this ecosystem. **Create** a diagram to model how energy flows. Your diagram should show how energy enters your ecosystem and how it moves from organism to organism. Your diagram should account for all the energy that enters your ecosystem to show how it moves and where it goes as it moves through your ecosystem.

Understanding Relationships

Explain the diagram you chose to represent energy flow in your mini-ecosystem. **Describe** how you developed this diagram as it relates to energy and matter in an ecosystem.

Grade 5 Resources

- **Bubble Map**
- **Safety in the Science Classroom**
- **Vocabulary Flash Cards**
- **Glossary**
- **Index**

Name _____

Bubble Map

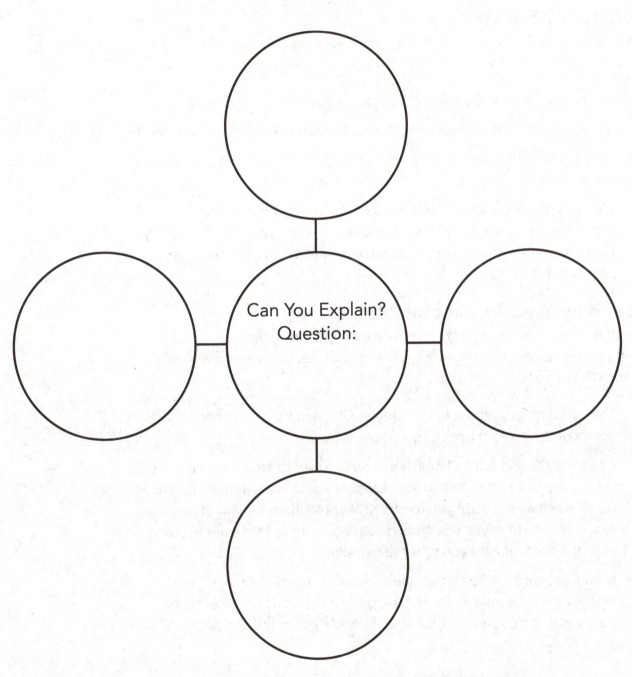

Bubble Map | R3

Safety

Safety in the Science Classroom

Following common safety practices is the first rule of any laboratory or field scientific investigation.

Dress for Safety

One of the most important steps in a safe investigation is dressing appropriately.

- Splash goggles need to be kept on during the entire investigation.
- Use gloves to protect your hands when handling chemicals or organisms.
- Tie back long hair to prevent it from coming in contact with chemicals or a heat source.
- Wear proper clothing and clothing protection. Roll up long sleeves, and if they are available, wear a lab coat or apron over your clothes. Always wear close toed shoes. During field investigations, wear long pants and long sleeves.

Be Prepared for Accidents

Even if you are practicing safe behavior during an investigation, accidents can happen. Learn the emergency equipment location in your classroom and how to use it.

- The eye and face wash station can help if a harmful substance or foreign object gets into your eyes or onto your face.
- Fire blankets and fire extinguishers can be used to smother and put out fires in the laboratory. Talk to your teacher about fire safety in the lab. He or she may not want you to directly handle the fire blanket and fire extinguisher. However, you should still know where these items are in case the teacher asks you to retrieve them.
- Most importantly, when an accident occurs, immediately alert your teacher and classmates. Do not try to keep the accident a secret or respond to it by yourself. Your teacher and classmates can help you.

Practice Safe Behavior

There are many ways to stay safe during a scientific investigation. You should always use safe and appropriate behavior before, during, and after your investigation.

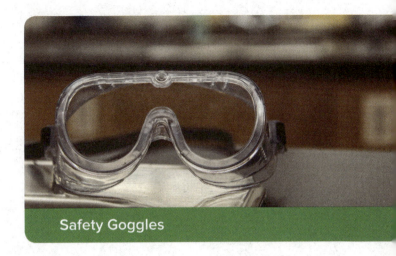
Safety Goggles

- Read the all of the steps of the procedure before beginning your investigation. Make sure you understand all the steps. Ask your teacher for help if you do not understand any part of the procedure.

- Gather all your materials and keep your workstation neat and organized. Label any chemicals you are using.

- During the investigation, be sure to follow the steps of the procedure exactly. Use only directions and materials that have been approved by your teacher.

- Eating and drinking are not allowed during an investigation. If asked to observe the odor of a substance, do so using the correct procedure known as wafting, in which you cup your hand over the container holding the substance and gently wave enough air toward your face to make sense of the smell.

- When performing investigations, stay focused on the steps of the procedure and your behavior during the investigation. During investigations, there are many materials and equipment that can cause injuries.

- Treat animals and plants with respect during an investigation.

- After the investigation is over, appropriately dispose of any chemicals or other materials that you have used. Ask your teacher if you are unsure of how to dispose of anything.

- Make sure that you have returned any extra materials and pieces of equipment to the correct storage space.

- Leave your workstation clean and neat. Wash your hands thoroughly.

Safety in the Science Classroom | R5

Vocabulary Flash Cards

absorb

Image: Bengaltigerdzr / Shutterstock.com

to take in

air

Image: Discovery Communications, Inc.

the part of the atmosphere closest to Earth; the part of the atmosphere that organisms on Earth use for respiration

chemical energy

Image: Discovery Communications, Inc.

the energy that is stored in the bonds between atoms

consumer

Image: Paul Fuqua

an organism that eats other living things to get energy; an organism that does not produce its own food

Vocabulary Flash Cards | R7

cycle

Image: inbevel / Shutterstock.com

a process that repeats

decomposer

Image: Paul Fuqua

organisms which carry out the process of decomposition by breaking down dead or decaying organisms

ecosystem

Image: Paul Fuqua

all the living and nonliving things in an area that interact with each other

energy

Image: Paul Fuqua

the ability to do work or cause change; the ability to move an object some distance

Vocabulary Flash Cards | R9

energy pyramid

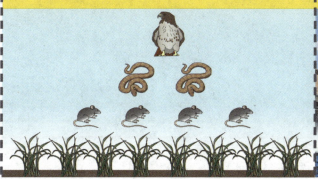

a model that shows the available amount of energy in each trophic layer in an ecosystem

energy transfer

the transfer of energy from one organism to another; or the transfer of energy from one object to another

environment

all the living and nonliving things that surround an organism

food chain

a model that shows one set of feeding relationships among living things

food web

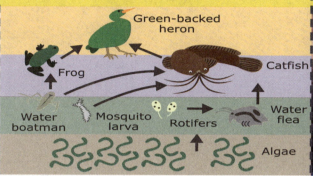

a model that shows many different feeding relationships among living things

interact

to act on one another

light

waves of electromagnetic energy; electromagnetic energy that people can see

light energy

a form of energy that animals can see directly; visible electromagnetic radiation

mechanical energy

energy that an object has because of its motion or its position

nuclear energy

the energy released when the nucleus of an atom is split apart (during fission) or combined with another nucleus (during fusion)

nucleus

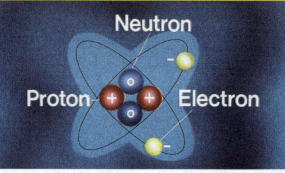

the center of an atom containing protons and neutrons

nutrients

important particles found in food that a living thing needs to survive

oxygen

Image: Paul Fuqua

a gas in Earth's atmosphere and in water that living organisms breathe

plant

an organism that is made up of many cells, makes its own food through photosynthesis, and cannot move

Image: Paul Fuqua

prey

Image: Discovery Communications, Inc.

an animal that is hunted and eaten by another animal

produce (v)

Image: panpote/Shutterstock

to make or create something

Vocabulary Flash Cards | R17

producer

Image: Paul Fuqua

an organism that makes its own food; an organism that does not consume other plants or animals

radiant energy

energy that does not need matter to travel; light

radiation

Image: Discovery Communications, Inc.

electromagnetic energy

recycle

Image: Rawpixel.com/Shutterstock

to create new materials from used products

Vocabulary Flash Cards | R19

shelter

Image: Discovery Education

(n.) a location or structure that gives protection (v.) to protect a person or thing from danger or bad weather

sound

Image: Paul Fuqua

a vibration that travels through a material, such as air or water; something that you sense with your hearing

stem

Image: Comstock Images / Stockbyte / Getty Images

the part of a plant that grows away from the roots; supports leaves and flowers

stomata

Image: Goodshoot / Getty Images

pores on the surface of a plant that allow gases to move into and out of the plant

Vocabulary Flash Cards | R21

substance

Image: Olga Chesnokova/Shutterstock

the physical matter of which living or nonliving things are composed

sun

Image: Paul Fuqua

any star around which planets revolve

survive

Image: Monkey Business Images / Shutterstock.com

to continue living or existing: An organism survives until it dies; a species survives until it becomes extinct.

system

Image: Gil Meshulam/Shutterstock

a group of related objects that work together to perform a function

terrarium

Image: Pixabay

a man-made ecosystem used for organisms to grow, usually in a clear container to allow for observation

water

Image: Paul Fuqua

a compound made of hydrogen and oxygen

Vocabulary Flash Cards

Glossary

English ——— A ——— Español

abiotic
includes all nonliving things

abiótico
incluye todos los objetos sin vida

absorb
to take in

absorber
incorporar

abyssal zone
the zone of deep ocean below a depth of 4,000 meters

zona abisal
la zona del océano profundo por debajo de los 4,000 metros de profundidad

air
the part of the atmosphere closest to Earth; the part of the atmosphere that organisms on Earth use for respiration

aire
parte de la atmósfera más cercana a la Tierra; la parte de la atmósfera que los organismos que habitan la Tierra utilizan para respirar

aquatic
relating to water

acuático
relativo al agua

Arctic
being from an icy climate, such as the North Pole

ártico
que tiene relación con el Polo Norte o el área que lo rodea

astronaut
a person who travels outside Earth's atmosphere; a person who travels into space (related words: astronomer, astronomy)

astronauta
persona que viaja fuera de la atmósfera de la Tierra; una persona que viaja por el espacio (palabras relacionadas: astrónomo, astronomía)

atmosphere
layers of gas that surround a planet (related word: atmospheric)

atmósfera
capas de gas que rodean un planeta (palabra relacionada: atmosférico)

axis
an imaginary line that an object spins or revolves around

eje
línea imaginaria sobre la cual gira o da vueltas un objeto

---— B ———

beaker
a scientific measuring cup used to measure liquids

vaso de precipitados
contenedor cilíndrico de vidrio usado en laboratorio, frecuentemente marcado con líneas para indicar capacidad

biodiversity
all of the organisms that live together in an environment

biodiversidad
todos los organismos que viven juntos en un medio ambiente

biosphere
that part of Earth in which life can exist

biosfera
parte de la Tierra donde puede existir la vida

biotic
includes all living things

biótico
incluye a todos los seres vivos

boil
to change state from a liquid to a gas because of added heat

hervir
cambiar el estado de líquido a gas por el aumento de la temperatura

---

carbon dioxide
a waste product made by cells of the body; a gas in the air made of carbon and oxygen atoms: humans rid themselves of carbon dioxide waste by exhaling, or breathing out

dióxido de carbono
desecho formado por células del cuerpo; gas del aire formado por átomos de carbono y oxígeno: los humanos desechamos dióxido de carbono al exhalar o expulsar aire

celestial sphere

an imaginary sphere surrounding Earth that show the sky and stars overhead

esfera celeste

gran esfera imaginaria centrada en un observador y que abarca el cielo encima de él

Celsius

the metric temperature scale: Water boils at 100 degrees Celsius, and it freezes at 0 degrees Celsius

Celsius

escala de la temperatura métrica: el agua hierve a los 100 grados centígrados y se congela a los 0 grados centígrados

change of state

the process of changing (by heating or cooling) one form of matter into another form of matter with different characteristics

cambio de estado

el proceso de cambiar (mediante frío o calor) una forma de materia a otra forma de materia con diferentes características

chemical change

a process that changes substances into new substances and is different physically from the original substance

cambio químico

reacción química; proceso que transforma a las sustancias en nuevas sustancias (palabra relacionada: reacción química)

chemical energy
energy that can be changed into motion and heat, stored in the bonds between atoms

energía química
energía que está almacenada en las cadenas entre átomos

combine
to bring together; to mix (related word: combination)

combinar
unir, mezclar (palabra relacionada: combinación)

conserve
to protect something or prevent the wasteful overuse of a resource

conservar
proteger algo o evitar el uso excesivo e ineficiente de un recurso

constant
continuing without interruption

constante
que continúa sin interrupción

constellation
a particular area of the sky; a group of stars

constelación
área particular del cielo; grupo de estrellas

consumer
an organism that eats other living things to get energy; an organism that does not produce its own food (related word: consume)

consumidor
organismo que come otros seres vivos para obtener energía; organismo que no produce su propio alimento (palabra relacionada: consumir)

crater
a large, circular pit in the surface of a planet or other body in space usually formed when two bodies in space collide

cráter
hoyo grande y circular en la superficie de un planeta u otro cuerpo en el espacio, generalmente formado cuando chocan dos cuerpos en el espacio

cycle
a process that repeats (related word: cyclic)

ciclo
proceso que se repite (palabra relacionada: cíclico)

— D —

decomposer
organism that carries out the process of breaking things down into dead or decaying organisms

descomponedor
organismo que lleva a cabo el proceso de descomposición mediante la desintegración de los organismos muertos

E

Earth
the third planet from the sun; the planet on which we live (related words: earthly; earth – meaning soil or dirt)

Tierra
tercer planeta desde el Sol; planeta en el cual vivimos (palabras relacionadas: terrenal; tierra en el sentido de suelo o suciedad)

ecosystem
all the living and nonliving things in an area that interact with one another

ecosistema
todos los seres vivos y objetos sin vida de un área, que se interrelacionan entre sí

electron
a particle with a negative charge

electrón
partícula subatómica con una carga negativa

energy
the ability to do work or cause change; the ability to move an object some distance

energía
habilidad para hacer un trabajo o producir un cambio; habilidad para mover un objeto a cierta distancia

energy pyramid
a model that represents energy transfer within an ecosystem, showing a lot of energy at the bottom and a much smaller amount on top

pirámide energética
un modelo que representa la transferencia de energía dentro de un ecosistema, que muestra mucha energía en la base y una cantidad mucho más pequeña en la cima

energy transfer
the transfer of energy from one organism to another through a food chain or web; or the transfer of energy from one object to another, such as heat energy

transferencia de energía
transmisión de energía desde un organismo a otro a través de una cadena o red de alimentos; o transferencia de energía desde un objeto a otro, como por ejemplo la energía del calor

environment
all the living and nonliving things that surround an organism

medio ambiente
todos los seres vivos y objetos sin vida que rodean a un organismo

estuary
a coastal body of water where freshwater from a river mixes with saltwater from the ocean

estuario
cuerpo costero de agua donde el agua dulce de un río se mezcla con el agua salada del océano

— F —

food chain
a model that shows how energy passes from one organism to another in an ecosystem

cadena alimentaria
modelo que muestra un conjunto de relaciones de alimentación entre seres vivos

food web
shows the ways in which many food chains work with one another in an ecosystem

red alimentaria
modelo que muestra muchas y diferentes relaciones de alimentación entre los seres vivos

force
a pull or push that is applied to an object

fuerza
acción de atraer o empujar que se aplica a un objeto

freshwater
water that is not salty, such as that found in streams and lakes

agua dulce
agua que no es salada, como por ejemplo la que se encuentra en arroyos y lagos

friction
a force that stops motion

fricción
fuerza que se opone al movimiento de un cuerpo sobre una superficie o a través de un gas o un líquido

--- G ---

galaxy
a group of solar systems, dust, and gas held together by gravity; our solar system is part of the Milky Way galaxy

galaxia
grupo de sistemas solares, polvo y gas unidos por la gravedad; nuestro sistema solar es parte de la galaxia llamada Vía Láctea

gas
a state of matter without any defined volume or shape (related word: gaseous)

gas
estado de la materia sin volumen ni forma definidos en el cual los átomos o moléculas se mueven casi libremente (palabra relacionada: gaseoso)

geosphere
Earth's crust, both beneath the oceans and continents, as well as the mantle and inner and outer core

geosfera
corteza terrestre, tanto debajo de los océanos como de los continentes, así como también el manto y los núcleos interior y exterior

glacier
a large sheet of ice or snow that moves slowly over Earth's surface

glaciar
ran sábana de hielo o nieve que se mueve lentamente sobre la superficie de la Tierra

gravity
the force that pulls an object toward the center of Earth (related word: gravitational)

gravedad
fuerza que existe entre dos objetos cualquiera que tienen masa (palabra relacionada: gravitacional)

---- H ----

heat
the transfer of thermal energy

calor
transferencia de energía térmica

horizon
the point at which Earth's surface appears to meet the sky

horizonte
punto en el cual la superficie de la Tierra parece reunirse con el cielo

hydrogen
the most abundant element in the universe, made of one proton and one electron

hidrógeno
elemento químico que consiste en uno o más protones y un electrón

hydrosphere
all of the water on, under, and above Earth

hidrósfera
toda el agua que se encuentra sobre, debajo y en la Tierra

--- I ---

imaginary
existing only in your mind or in your imagination

imaginario
que existe sólo en nuestra mente o en nuestra imaginación

interact
to act on one another (related word: interaction)

interactuar
ejercer influencia mutua (palabra relacionada: interacción)

--- L ---

light
a form of energy that moves in waves and particles and can be seen

luz
ondas de energía electromagnéticas; energía electromagnética que la gente puede ver

light energy
that form of energy that animals can see directly; visible electromagnetic radiation

energía lumínica
tipo de energía que los animales pueden ver directamente; radiación electromagnética visible

light year
the distance light travels in a vacuum in one year; about 6 trillion miles

año luz
distancia que viaja la luz en el espacio en un año; alrededor de 6 billones de millas

liquid
a state of matter with a defined volume but no defined shape

líquido
estado de la materia que posee un volumen definido, pero no una forma definida

location
a place where something is positioned

ubicación
un lugar donde se posiciona algo

M

magnet
an object with a north and south pole that produces a magnetic field (related terms: magnetism, magnetic)

imán
objeto con un polo norte y un polo sur que produce un campo magnético (palabras relacionadas: magnetismo, magnético)

magnify
to make something appear larger, usually by using one or more lenses

ampliar
hacer que algo parezca más grande, generalmente usando una o más lentes

mass
the amount of matter in an object

masa
cantidad de materia en un objeto

material
things that can be used to build or create something

material
cosas que se pueden usar para construir o crear algo

matter
material that has mass and takes up some amount of space

materia
material que tiene masa y ocupa cierta cantidad de espacio

measure
to use a tool to learn more about the volume, length, or weight of an object (related word: measurement)

medir
usar una herramienta para saber más sobre el volumen, la longitud, o el peso de un objeto (palabra relacionada: medición)

mechanical energy
energy that an object has because of its motion or its position

energía mecánica
energía que tiene un objeto debido a su movimiento o posición

melt
to change a substance from solid to liquid

derretirse
cambio de sustancia de estado sólido a líquido

mineral
a natural, solid substance found in rocks; each mineral has a specific chemical makeup

mineral
sustancia natural y sólida que se encuentra en las rocas; cada mineral tiene una composición química específica

mixture
a combination of substances that can be physically separated from one another

mezcla
combinación de sustancias que pueden separarse físicamente unas de otras

model
a drawing, object, or idea that represents a real event, object, or process

modelo
dibujo, objeto o idea que representa un evento, objeto, o proceso real

Glossary | R41

moon
a body in outer space that orbits a planet; a natural satellite

luna
cuerpo en el espacio exterior que gira alrededor de un planeta; satélite natural

motion
a change in the position of an object compared to another object (related terms: move, movement)

movimiento
cambio en la posición de un objeto en comparación con otro objeto (palabras relacionadas: mover, desplazamiento)

--- N ---

natural resources
resources that are obtained from Earth

recursos naturales
recursos obtenidos de la tierra

nebula
an interstellar cloud made up of hydrogen gas, plasma, helium gas, and dust

nebulosa
una nube interestelar constituida por gases (hidrógeno y helio), plasma, y polvo

nonrenewable
once it is used, it cannot be made or reused again

no renovable
no renovable

nonrenewable resource
a natural resource of which a finite amount exists, or one that cannot be replaced with currently available technologies

recurso no renovable
recurso natural del cual existe una cantidad finita, o uno que no puede remplazarse con las tecnologías actualmente disponibles

nuclear energy
the energy released when the nucleus of an atom is split apart (during fission) or combined with another nucleus (during fusion)

energía nuclear
energía liberada cuando el núcleo de un átomo se divide (durante la fisión) o combina con otro átomo (durante la fusión)

nucleus
the center of an atom containing protons and neutrons (related terms: nuclei, nuclear); a region in a cell that is surrounded by a membrane and contains genetic material

núcleo
centro de un átomo que contiene protones y neutrones (palabras relacionadas: núcleos, nuclear); región de una célula que está rodeada por una membrana y contiene material genético

nutrients
important particles found in food that a living thing needs to survive

nutrientes
sustancia como la grasa, una proteína o un carbohidrato que un ser vivo necesita para sobrevivir

O

ocean
a large body of saltwater that covers most of Earth

océano
gran cuerpo de agua salada que cubre la mayor parte de la Tierra

optical
having to do with the eye or lenses

óptico
relacionado con los ojos o con las lentes

orbit
the circular path of an object as it revolves around another object

órbita
trayectoria circular de un objeto que se forma a medida que gira alrededor de otro objeto

organism
any individual living thing

organismo
todo ser vivo individual

oxygen
a gas found in air and water that living things need to breathe

oxígeno
gas que se encuentra en la atmósfera de la Tierra y en el agua que los organismos vivos respiran

P

particle
something that is very tiny

partícula
algo que es muy pequeño

perpendicular
a downward direction, making a right angle

perpendicular
con dirección descendente, formando un ángulo recto

phenomena
a scientific wonder or happening

fenómeno
una maravilla o suceso científico

physical change
a change in matter that does not change the makeup of a substance

cambio físico
alteración en la materia que no afecta su composición química

plankton
small organisms that drift through bodies of water; include animals, plants, and bacteria

plancton
organismos pequeños que se mueven sin rumbo a través de cuerpos de agua; incluyen animales, plantas, y bacterias

plant
an organism that is made up of many cells, makes its own food through photosynthesis, and cannot move; a member of kingdom Plantae

planta
organismo formado por muchas células que fabrica su propio alimento a través de la fotosíntesis, y no se puede mover; miembro del reino vegetal

plasma
a fourth state of matter in which the particles of a gas become highly charged (ionized)

plasma
un cuarto estado de la materia en el que las partículas de un gas se vuelven muy cargadas (ionizadas)

pollute
to put harmful materials into the air, water, or soil (related words: pollution, pollutant)

contaminar
poner materiales perjudiciales en el aire, agua, o suelo (palabras relacionadas: contaminación, contaminante)

precipitation
water that is released from clouds in the sky; includes rain, snow, sleet, hail, and freezing rain

precipitación
agua liberada de las nubes en el cielo; incluye la lluvia, la nieve, la aguanieve, el granizo, y la lluvia congelada

prey
an animal that is hunted and eaten by another animal

presa
animal que es cazado y comido por otro

produce (v)
to make or create something

producir (v)
hacer o crear algo

producer
an organism that makes its own food; an organism that does not consume other plants or animals

productor
organismo que fabrica su propio alimento; organismo que no consume otras plantas u otros animales

property
a characteristic or quality of a material

propiedad
una característica o calidad de un material

R

radiant energy
energy that does not need matter to travel; light

energía radiante
energía que no necesita de la materia para viajar; luz

radiation
heat or light energy (related word: radiate)

radiación
energía electromagnética (palabra relacionada: irradiar)

recycle
to create new materials from used products

refuse (n)
garbage

relies
dependent upon

renewable resource
a natural resource that can be replaced

resource
a naturally occurring material in or on Earth's crust or atmosphere of potential use to humans

revolution
the orbiting of an object around another object

reciclar
crear nuevos materiales a partir de productos usados

desecho (s)
basura

depender
estar sujeto a, estar subordinado a

recurso renovable
recurso natural que puede reemplazarse

recurso
material que se origina de forma natural en o sobre la corteza o la atmósfera de la Tierra, que es de uso potencial para los seres humanos

revolución
movimiento por el cual un objeto gira alrededor de otro objeto describiendo una órbita completa

river
big body of water flowing through land on either side

río
gran cuerpo de agua que fluye con tierra de ambos lados

rotate
turning around on an axis; spinning (related word: rotation)

rotar
girar sobre un eje; dar vueltas (palabra relacionada: rotación)

rotation
the spinning of a celestial body, such as a planet, around an axis

rotación
giro de un cuerpo celeste, como un planeta, alrededor de un eje

--- S ---

salt
a mineral found in the ocean and other parts of Earth that can be used for preserving things and seasoning food

sal
compuesto iónico formado por iones de carga positiva y por iones de carga negativa, por lo tanto el producto no tiene carga

saltwater
contains salt and other minerals that make the water unsuitable for drinking

agua salada
término general que designa agua que contiene más de 1 parte por millar (1 g/L) de sólidos disueltos totales

scale
a device used for measuring weight

escala/báscula
descripción del tamaño o de la cantidad relativa de dos o más cosas; marcas de una herramienta de medición que representan unidades particulares; aparato utilizado para medir el peso

shelter
a location or structure that gives protection from danger or bad weather

refugio
un lugar o estructura que brinda protección a una persona u objeto contra los peligros o las malas condiciones meteorológicas

solar system
a system of objects that revolve around a star

sistema solar
conjunto de objetos que giran alrededor de una estrella

solid
matter with a fixed volume and shape

sólido
materia con un volumen y una forma determinada

sound
anything you can hear that travels by making vibrations in air, water, and solids

sonido
vibración que viaja a través de un material, como el aire o el agua; lo que se percibe a través de la audición

star
a massive ball of gas in outer space that gives off heat, light, and other forms of radiation

estrella
bola masiva de gas en el espacio exterior que emite calor, luz, y otras formas de radiación

state of matter
a particular form that matter can take: the three main states of matter are solid, liquid, and gas.

estado de la materia
forma particular que puede tener la materia: los tres estados principales de la materia son sólido, líquido, y gaseoso

stem
the part of a plant that grows away from the roots; supports leaves and flowers

tallo
parte de la planta que crece a partir de la raíz; contiene las hojas y la flores

stomata
pores on the surface of a plant that allow gases to move into and out of the plant (related word: stoma)

estomas
poros en la superficie de una planta que permiten a los gases moverse hacia adentro y hacia afuera de la planta (palabra relacionada: estoma)

stream
a small body of flowing water

arroyo
pequeño cuerpo de agua que fluye

substance
the physical matter of which living or nonliving things are made

sustancia
materia física de la cual están compuestos los seres vivos y objetos sin vida

sugar
a substance that is sweet, mainly used in food and drinks

azúcar
compuesto químico usado por los organismos para obtener energía

sun
any star around which planets revolve

sol
toda estrella alrededor de la cual giran los planetas

surface
the top or outermost part of something

superficie
parte superior de un objeto; exterior de un objeto; límite entre dos objetos o materiales

survive
to continue living or existing: an organism survives until it dies; a species survives until it becomes extinct (related word: survival)

sobrevivir
continuar viviendo o existiendo: un organismo sobrevive hasta que muere; una especie sobrevive hasta que se extingue (palabra relacionada: supervivencia)

sustainable
able to be used over and over again without hurting the overall supply

sostenible
que se puede utilizar una y otra vez sin afectar el suministro total

system
a group of related objects that work together to perform a function

sistema
grupo de objetos relacionados que funcionan juntos para realizar una función

T

telescope
an instrument used to observe objects that are far away

telescopio
instrumento usado para observar objetos que se encuentran alejados

terrarium
a human-made ecosystem used for organisms to grow, usually in a clear container to allow for observation

terrario
un ecosistema creado por el hombre para que crezcan organismos, generalmente en un recipiente transparente para permitir la observación

thermal energy
energy in the form of heat

energía térmica
energía en forma de calor

U

universe
everything that exists in, on, and around Earth

universo
todo lo que existe en, sobre, o alrededor de la Tierra

volcano
an opening in Earth's surface through which magma and gases or only gases erupt (related word: volcanic)

volcán
abertura en la superficie de la Tierra a través de la cual el magma y los gases o sólo los gases hacen erupción (palabra relacionada: volcánico)

volume
the amount of space that matter takes up

volumen
la cantidad de espacio que ocupa la materia

warm
having heat

cálido
que tiene calor

water
a compound made of hydrogen and oxygen; can be in either a liquid, ice, or vapor form and has no taste or smell

agua
compuesto formado por hidrógeno y oxígeno

watershed
a region in which all precipitation and surface water collects and drains into the same river

cuenca
una región en la que se recoge toda la precipitación y agua superficial, que drena hacia el mismo río

weight
the force of gravity on an object

peso
fuerza de gravedad que se ejerce sobre un objeto

Index

A

Absorb 66
Air 26, 36, 87–88, 118
Analyze Like a Scientist 26–28, 36–37, 43, 54–55, 65–66, 73–74, 76–77, 82–83, 87–89, 98–101, 116–119, 123
Animal 54, 58, 66, 88, 98
Ask Questions Like a Scientist 10–11, 52–53, 108–109

B

Bat 98
Bee 98–99
Bird 98–99
Butterfly 98–99

C

Calorie 135
Can You Explain? 8, 41, 50, 95, 106, 131
Chemical energy 116
Consumer 76, 83, 134
Cycle 83

D

Decay 56
Decomposer 60, 82–84, 86
Desert 13, 139

E

Ecosystem 56–59, 73, 83, 88
Energy
    as food 36, 65–66, 73, 76, 88, 110, 114–118, 135
    flow 73, 115, 125, 134
    pyramid 123–124
    transfer 117–118, 123
Environment 66, 83
Evaluate Like a Scientist 15–17, 35, 39, 46, 57–59, 75, 80–81, 92, 102–103, 110–112, 114, 121–122, 124, 128, 137

F

Food
    properties 17, 26, 36–38, 52, 65–66, 94
    chain 72–76, 118, 123, 134
    web 76–80, 128, 134

G

Glucose 36, 66

H

Habitat 99
Hands-On Activities 18–21, 22–25, 30–32, 60–64

I

Interact 76, 88, 139
Investigate Like a Scientist 18–25, 30–32, 60–64
Irrigation 43

L

Leaves 36–39, 66, 88
Light energy 36, 116

M

Mechanical energy 117

N

Nuclear energy 117
Nucleus 117
Nutrient 26, 36, 83, 88

O

Observe Like a Scientist 12–14, 29, 33–34, 38, 56, 72, 78–79, 84–86, 90–91, 120
Oxygen 36, 65

P

Photosynthesis 44
Plants
    as producers 36, 76
    decomposition of 83
    growing 43–44, 66, 88
    properties of 26
Prey 73
Producer 76, 83, 134

R

Radiant energy 116
Rain 43
Record Evidence Like a Scientist 40–42, 94–97, 130–133
Recycle 83
Root 26, 88

S

Scavenger 82
Shelter 88
Soil 18, 26, 88
Solve Problems Like a Scientist 4–5, 138–143
Sound 116
Stem 26, 30
STEM in Action 43–45, 98, 134–136
Stomata 26
Substance 88
Sun
    as light energy 36, 66, 73, 76, 116, 123
    in growing plants 26, 36, 43–44, 88
Survive 26
System 26, 39, 120

T

Terrarium 139
Thermal energy 116
Think Like a Scientist 67–71, 115, 125–127
Tree 10, 40

U

Unit Project 4–5, 138–143

V

Vessel 26, 36

W

Waste 36, 83
Water
    as part of photosynthesis 26, 36, 66
    in irrigation 43–44